WEED

T&J

Published by TAJ Books International LLC 2013
219 Great Lake Drive
Cary, North Carolina, USA
27519
www.tajbooks.com

All notations of errors or omissions (author inquiries, permissions) concerning the content of this book should be addressed to info@tajbooks.com.

ISBN 978-1-84406-244-7

Printed in China.
1 2 3 4 5 17 16 15 14 13

WEED

BY JO FORTY

INTRODUCTION

Cannabis (aka weed, marijuana, ganga) is the product of the cannabis plant prepared for use as a psychoactive drug and as a medicine. Its use stretches as far back as history itself. One of the earliest known plants to be domesticated by man, human knowledge of the plant's practical and recreational uses has developed over many years. This plant has so much potential: it grows easily almost anywhere; from its fibrous stems hemp can be made and thus rope, paper, cloth, and many other products; from its seeds oil is produced with a variety of uses; and from its flowers comes the controversial drug....

Cannabis, or weed, is the fourth most widely used drug in the world after coffee, alcohol, and tobacco. The United Nations' estimate for global consumption of cannabis in 2010 was 2.6–5.0% of the adult population—between 119 million and 224 million users—although this statistic does not include all likely users in Asia, the plant's place of origin and the location of its greatest cultural acceptance.

But unlike coffee, alcohol, and tobacco, cannabis is illegal in most of the world and is, therefore, the world's most widely used illicit drug. This has led to the current complex legal position as regards the drug, encompassing a range of attitudes toward its use as well as toward punishments and sentencing. The lack of any serious toxic or addictive properties that would justify such a strongly antipathetic opinion has led to the development of the "gateway drug" theory: cannabis use leads people to try other drugs. To cloud things still further, increasingly undeniable evidence exists of its beneficial medicinal properties in treating various ailments and conditions. A huge expenditure of manpower and resources incurred by the "war on drugs," a predominantly Western war, continues unabated. The huge resources that have already been committed, combined with the vested interests of governments, drug cartels, and terrorist movements that have moved into drugs as a way of financing their operations, now ensures that the war will be endless.

All that being said, an important 2012 international study in New Zealand showed that cannabis use by younger people needs to be approached carefully. Those who used it regularly below the age of 18—while their brains were still developing—suffered a noticeable drop in their intelligence quotient, or IQ. "It is such a special study that I'm fairly confident that cannabis is safe for over-18 brains, but risky for under-18 brains," said Professor Terrie Moffitt, Institute of Psychiatry, King's College London.

Macro cannabis bud

Cannabis has a pungent and distinctive odor and is smoked mixed with tobacco as a cigarette or in a pipe. It is also smoked in "blunts": cigars that have been emptied and refilled with a mixture of marijuana and tobacco. These methods combine marijuana's active ingredients with nicotine. It can be smoked pure in all the ways just mentioned as well as vaporized. It can be cooked, mixed into food, or brewed as a tea. Its concentrated, resinous form is called hashish; the sticky black liquid is hash oil.

Opposite page: Top: THC-encrusted female flowers
Below left: Marijuana seeds
Below right: Loose marijuana

This page: Below: Cheap grinder
Right: Various kinds of hashish
Below right: Loaded blunt

Left: Pipes—a surreptitious modern composite bomb and a more traditional Asian wood and brass type

Below left: A Moroccan "Caramello" pellet of hashish—conveniently sized and shaped for body smuggling.

Right: Skunk bud nearing perfection, close to harvest

THE PLANT

The cannabis genus is a member of the nettle family and consists of three *taxa,* or types: *Cannabis sativa, C. indica,* and *C. ruderalis.* All are native to central and south Asia. Cannabis is an annual flowering plant and primarily dioecious (consisting of individual males and females). Its leaves are defined as palmately compound or digitate–like a hand, with serrated leaflet fingers. *C. sativa* grows tallest of the three, branched with relatively narrow leaflets; *C. indica* grows shorter and bushier with slightly wider leaflets; and *C. ruderalis* is short and branchless and grows wild in the Asian heartland of the plant's origin.

All types are intrinsically the same, but grow to different heights, have different uses for man, and each possesses different levels of THC (tetrahydrocannabinol)–the primary psychoactive element of the plant. Over time, different strains have been cultivated from the more useful sativa and indica types and grown according to the end requirement: low-level THC plants for their fiber and high-level plants for their psychoactive properties.

Today, C. sativa is primarily grown for industrial use (but can still have a substantial psychoactive impact), whereas C. indica, which has a poorer fiber quality, is used for recreational and medicinal purposes. Now, 2,000 strains or hybrids of the two *taxa* exist.

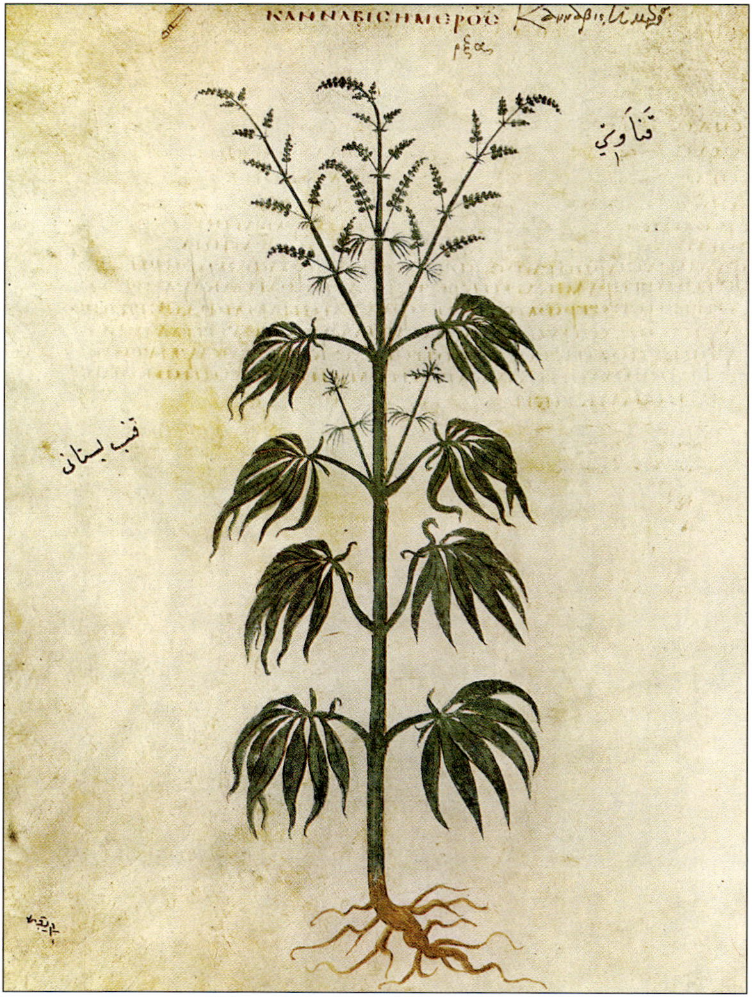

Left: Illustration of C. Sativa *from the early-6th-century illuminated manuscript of* De Materia Medica *by Dioscorides–one of the earliest records of cannabis in western sources*

Right: C. Sativa *from Prof. Dr. Otto Wilhelm Thomé's* Flora von Deutschland, Osterreich, und der Schweiz, *1885. Hemp was grown in many European countries and colonies as a cordage fiber. Its natural resistance to rot recommended it especially for maritime uses, and as European seafaring expanded, so too did its production.*

Cannabineae.
A
B
1
2
3
4
5
6
7
8
9
10
11
12
13
Cannabis sativa L.
W. Müller.

Above: The paired pistils (pollen-catching appendages) of a young female plant before flowering

Left: Cured "Sweet Tooth" bud, ready for smoking

Right: Close-up of a mature female bud, its trichomes (hairy leaflets) beginning to be covered in sticky globules of THC

Above: The female flower of an outdoor-grown C. indica

Right: The distinctive palmate cannabis leaf

Opposite page: The bulging pollen sacks of a heavily flowering male marijuana plant

Left and right: Cannabis male flower pollen sacks. When a male plant enters the stage of flower development, the tips of its branches start to grow what looks like a little bud, but without the female hairs. Most growers try to identify and remove male marijuana plants as soon as possible.

Left and right: C. ruderalis *(wild marijuana) foliage. Wild marijuana is the hardier, stockier native variety grown in the northern Himalayas and southern states of the former Soviet Union, and other areas where hemp cultivation was once prevalent. It has a lower THC content than either* C. sativa *or* C. indica *and so has little recreational use. Its resistance to insects and disease, along with its early and short flowering period, make it appealing to breeders who want to integrate these advantages into their Indica and Sativa strains.*

Above: Cannabis has a fibrous sheath
Opposite page: Above left: Hemp rope
Below left: Hemp shirt
Right: Hemp ale

HEMP

The first use of cannabis was to make fiber cord, then rope and textiles. By slitting the stalk and scraping out the pith, then retting or soaking it, a pliable string is left that can be woven. Hemp is the natural, durable soft fiber produced from the sheath-like stalk of C. sativa plants, which can grow over 20 feet tall. Hemp's medicinal and recreational properties emerged through processing.

As a crop, weed is tough and consequently easy to manage, with a growing cycle of around 100 days—faster and with a higher yield per acre than many of man's other domesticated plants. Hemp also requires few pesticides and no herbicides because of its height and the density at which it can be planted, making it environmentally friendly.

Today, industrial hemp is used to make many products including paper, textiles, clothing, biodegradable plastics, insulation, cosmetics, foodstuffs, animal feed, and bio-fuel. Continuing a medical tradition stretching back millennia, hemp producers sell the seeds as a health food, rich in heart-healthy, essential fatty acids, amino acids, vitamins, and minerals. A dairy- and gluten-free milk made from the seeds is also produced.

Hemp is harvested for commercial purposes in over 30 countries. China is the world's leading producer. Ironically, more hemp is exported to the United States than to any other country, even though the U.S. government often cannot distinguish between psychoactive cannabis and the non-psychoactive variant used for industrial and commercial purposes, uprooting and destroying hemp that grows in the wild across much of America. The latent ambiguity of the genus has proven a dilemma for U.S. authorities.

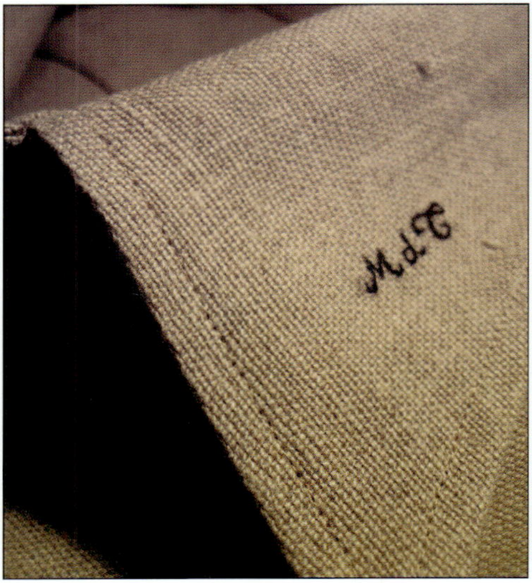

HEMP ALE
ALL NATURAL BROWN ALE
Family Operated
Since 1987

Almost a forest of hemp, which can grow over 20 feet tall
Opposite: Hemp seed

Above and right: Industrial C. sativa. *Tough and easy to manage, requiring few pesticides and no herbicides because of its height and the density at which it can be planted, hemp is an environmentally friendly crop used in a vast range of products.*

CANNABINOIDS

The complexity of this plant and its relationship with man surfaces again when we try to categorize it pharmacologically. Whereas most psychoactive drugs fall clearly into a single category–stimulant, depressant, or hallucinogen–cannabis cannot be so easily constrained, possessing as it does all these capabilities and more. The plant contains over 400 compounds including a unique group of 66 so-called cannabinoids, some of which are intensely psychoactive. These produce a wide range of moods and awareness in the user according to their ratios in different strains. The highest concentrations are found in the female flowers of the plant.

The four main cannabinoids are delta-9-tetrahydrocannabinol (delta-9-THC), cannabidiol (CBD), delta-8-tetrahydrocannabinol (delta-8-THC), and cannabinol (CBN). All are psychoactive apart

Below left: Delta-9-tetrahydrocannabinol-C4
Below: Raphael Mechoulam (R), discoverer of delta-9-THC from C. sativa *(1964) and Lumír Ond ej Hanu (L), discoverer of the endogenous ligand, anandamide, in the brain (1992). Both compounds bind to the cannabinoid receptors in the brain.*

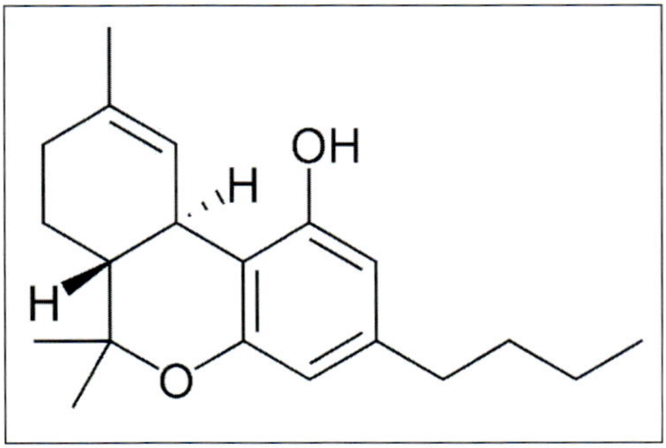

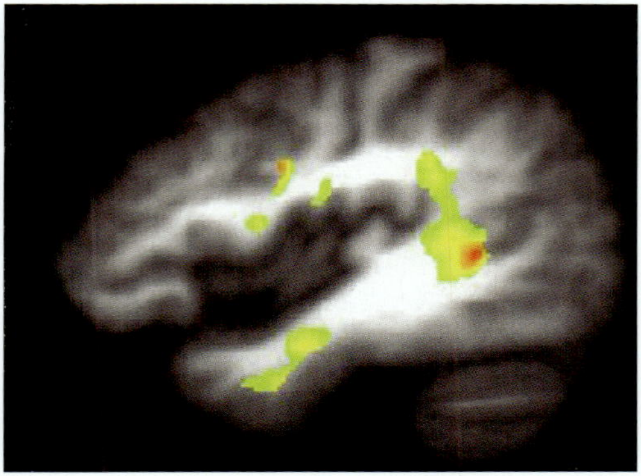

Left: "Spice"–a designer synthetic cannabinoid
Above: The activated areas of the brain of a marijuana user

from cannabidiol, which nevertheless interacts with the others to influence their effects. By far the most powerful is delta-9-THC.

Cannabinoids have such a strong effect on humans because we have cannabinoid-type receptors in our brains. The brain produces similar compounds itself, called endocannabinoids, which are used to control mood and thought. Most of these receptors are in the parts of the brain that deal with the senses, pleasure, memory, awareness, and time perception. Such receptors are also found in other parts of the body.

As our knowledge developed about cannabinoids and endocannabinoids, cannabis was increasingly viewed as a useful plant with a wide variety of applications: practical, spiritual, and medicinal. When smoked, THC enters the bloodstream instantaneously (when eaten it takes longer) and is transported to sites in the body and the brain where it binds to the cannabinoid receptors, usually releasing feelings of elation, euphoria, empathy, and sensitivity.

Just as in all life's processes, different factors and considerations influence the outcome. Some factors are internal, such as in the specific chemistry of a particular cannabis strain, and others are external. Each situation can bring forth its own specific responses.

The difficulty in classifying cannabis is also evinced in its effects, which can feature elements or symptoms of all other drugs. Sometimes weed can change the user's mood (for better or worse), and at other times it can magnify the user's existing mood. The intensity of feeling can also vary enormously, from subtle and suggestive to almost overwhelming. Perhaps one thing cannabis use does teach people is to never preconclude!

HISTORY

There is much evidence for the prolonged and consistent worldwide use of cannabis in ritual, religious, and medicinal contexts throughout recorded history. Following a recognizable pattern of human development, its use has evolved from the shaman/medicine man's magic trance rituals to the more organized rites of established religions and then to medicine, which is the accumulation of knowledge that creates substances mainly from plants, designed to heal, relieve, and re-balance the body. Thus, cannabis provides psychic and physical relief.

Across all continents, wherever man has ventured, these structures are apparent—sometimes in an unbroken continuity that proves the link beyond doubt. Found in the tomb of an ancient shaman in China, cannabis was used as a key to enter into dialog with the invisible world of the spirits. The spirit world was a manifestation and a viable explanation of the inexplicable, where man could attempt to deal with his deepest fears and needs. It explained where the ancestors had gone and who made the critical weather that could make or break crops. Through such ritual people sought reassurance.

Hand-in-hand with this psychoactive capability was the medicinal value of the plant as an analgesic, sedative, anti-inflammatory, antispasmodic, and anticonvulsant. Cannabis is first textually mentioned in China in the third millennium BC as a treatment for malaria

An ancient Turkish saying:
"*Helal ottur, gunah yoktur.*" (It's a holy weed that carries no sin.)

Turks smoking the narghile, *a traditional Eastern water-cooled pipe*

Left: The Nargileh *by Maurice Orange was shown at the 1906 Paris Salon.*

and rheumatism, although its intoxicating potential was also acknowledged.

In India, mother country of philosophies and religions, cannabis was used in meditation, religious rites, and medicine and is still so used today. In the Middle East, with the rise of Islam, cannabis evolved still further, being pressed into a solid resin to be smoked in water-cooled pipes. In Afghanistan, Lebanon, Indonesia, Egypt, Morocco, and Turkey, the Muslim prohibition on alcohol, along with its medicinal properties, spurred the acceptance and development of cannabis.

With the expansion of European colonies in Asia and the Middle East, westerners had closer contact with weed and its uses. In 1753, C. sativa was classified by Linneaus (father of modern binomial nomenclature), and in 1783 C. indica was classified by Lamarck. By the 1840s, medical papers were written on the benefits of cannabis in treating a variety of ailments. In that era, no controversy existed over its recreational use because it was only used by minority groups such as artists, musicians, and poets or natives from the countries where it was already culturally known and accepted.

Between man's first use of cannabis to the present time, legality was not an issue until about a century ago when the Western nation states, chiefly the U.S., took action.

Why?

Left: Ma Gu ("Hemp Maid") is a legendary Taoist Immortal in Chinese mythology. Cannabis has been continuously cultivated in China since Neolithic times, recognized for its practical, medicinal, and psychoactive qualities.

Above: Bhang shop, India. Used in countless ways since history began, cannabis is more accepted culturally in India than perhaps anywhere else in the world.

Far left: In Lebanon the hookah, or smoking pipe, is known by the Arabic word arguileh
Left: Ali enjoying his hubble-bubble pipe

Above: Indoor-cultivated "Hindu Kush" skunk

MARIJUANA IN THE U.S.

Cannabis was first brought to the Americas in the 1540s by the Spanish, followed by the English soon afterwards in the early 1600s. For the next few hundred years, hemp was grown in the New World, often under direct order of the government. The English grew it exclusively as a source of fiber in their colonies. The Spanish, however, having enjoyed a closer relationship with the Muslim world, including several centuries under Muslim occupation, had more of an understanding of the plant's other capabilities. The clue to its origin in the Americas lies in the name "marijuana." The first documented evidence for its recreational use comes from Mexico, when during the revolution, Pancho Villa's men sang a song about smoking marijuana. Later, Mexico became the exclusive source of marijuana for the U.S.

In the early years in the U.S., along with opium and cocaine, cannabis-based patented tinctures were made to relieve many ailments; given that this was distinctly in the medical sphere, no controversy arose over its use. But by the 1910s, marijuana had begun to emerge from the underground culture of Hispanic and African-American communities as a recreational drug. This was helped by Prohibition in the 1920s, when more interracial cultural connections were made because of the draconian anti-alcohol legislation.

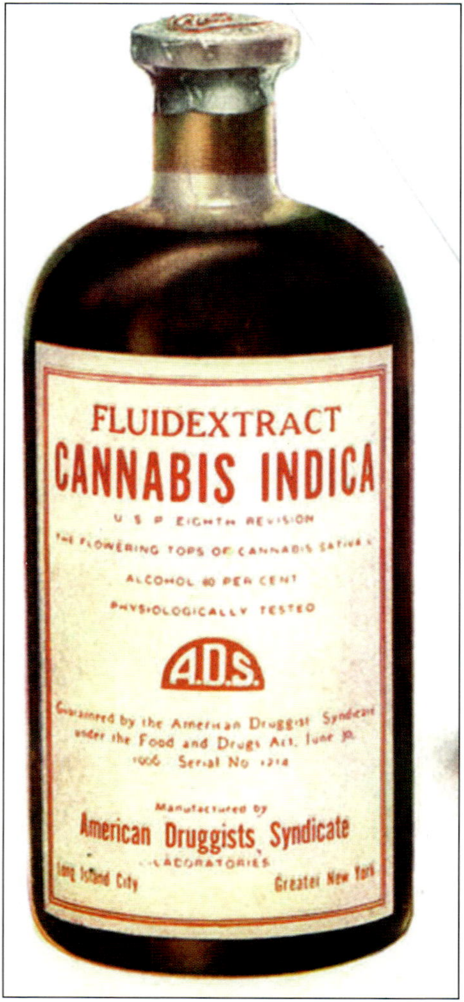

Left: Druggist's bottle containing medicinal
C. indica *tincture*
Above: Engraving of women processing hemp for the production of cordage

In the West the drug that most approximates cannabis is alcohol. An intrinsic part of ancient Greek and Roman culture, as in more modern cultures, alcohol has been used as a food, a drink (often purer than the available water because of its fermentation), a medicine, and in both religious and secular rites. Its potential for misuse was also recognized, but for most the benefits outweighed the harm and banning it completely was never considered. When the U.S. attempted alcohol prohibition in the 1920s, it was a dismal failure. By criminalizing so many normal people, the government securely established a criminal underworld that was viewed sympathetically by the public and caused problems that still exist to this day.

Spurred on by the outlawing of alcohol and the jazz age explosion of new musical and fashion styles, cannabis was glorified in "reefer songs," and clubs devoted to its use sprung up in most of the major U.S. cities. Listed in the U.S. pharmacopeia from 1850 until 1942 and prescribed for various ailments, it was still not illegal and not immediately considered a social threat by the authorities. But as time passed and its use as an intoxicant grew, the governmental authorities became more uncomfortable. In 1912, controls on cannabis were discussed at the first International Opium Conference. Various states began to make it illegal (New York, California, and Texas), and by 1924, at the second Opium Conference

Above: Titillating cannabis confessions for a prurient public

Right: Today a cult comedy, Reefer Madness *was originally sponsored by a religious group before being recut as an exploitation movie.*

TELL
YOUR
CHILDREN

32

it was classified as a dangerous addictive narcotic.

In the 1930s, the Bureau of Narcotics was formed, with Harry J. Anslinger appointed as its first Commissioner, a post he would hold for over 30 years. Behind the scenes, various business magnates whose vested interests had been affected were also involved in what was to become a determined attack against the legality of cannabis. In Mexico, newspaper tycoon Randolph Hearst had 80,000 acres of timber seized by Pancho Villa and his army. In retaliation, Hearst's papers begin to run lurid stories, portraying Hispanics and African-Americans as frenzied beasts under the influence of marijuana, the "devil weed." Hearst also teamed up with DuPont to create a similarly sensational anti-marijuana campaign in order to eliminate hemp as an industrial competitor.

By 1934, Anslinger was broadcasting on the radio in a similarly racist, alarmist style in his bid to outlaw the drug and raise hysteria from a state to a national level. The Marijuana

Right: A sizeable "field" of outdoor weed cultivated by a guerrilla grower

Far right: A cartoon about subverting Prohibition by Louis M. Glackens in Puck *magazine: "Said prohibition Maine to prohibition Georgia: 'Here's looking at you.'"*

GEORGIA
ORANGE PHOSPHATE
MAINE
COLD TEA
L. M. GLACKENS

Abraham Lincoln said in December 1840 that "Prohibition...goes beyond the bounds of reason in that it attempts to control man's appetite through legislation and makes a crime out of things that are not even crimes.... A prohibition law strikes a blow at the very principles upon which our Government was founded."

Tax Act of 1937 was passed and prohibited the production of hemp as well as cannabis.

Despite the legislation, however, cannabis use continued to increase. Postwar, it became a vital subcultural tool of the 1950s and 1960s. From the Beat Generation to the Yippies and Hippies it was an essential part of youthful rebellion and the dream of radical social change.

The Controlled Substances Act of 1970 classified marijuana, along with heroin and LSD, as a Schedule 1 drug with the highest abuse potential and no medical use. Mexico, the main U.S. supplier at that time, was persuaded to spray its crop with the herbicide Paraquat. The result was to move the cannabis industry to Colombia in South America, which became the next major U.S. supplier, and also encouraged the homegrown U.S. market.

The "zero tolerance" and "war on drugs" campaigns, accompanied by stricter laws, border controls, and harsh punishments initiated by the Reagan and Bush administrations were largely unsuccessful.

Left and this page: Operation "Pipeline Express" seized 61,573 pounds of cannabis in 2011, part of the continuing war against drugs.

MODERN METHODS OF CULTIVATION

As knowledge of all plants increases, along with improvements in horticultural skills, new types or cultivars of plants are created. Because of the cannabis plant's legal status and the government's intransigent attitude, people began to grow weed more secretly inside their own homes, which led to another series of developments.

More complicated and expensive than outdoor cultivation, but allowing the cultivator complete control of the growing environment and the plant itself, new internal growing methods were invented, using high-intensity lighting systems to mimic the sun, including hydroponics (a water-based growing medium), aeroponics (air-based growing), and sea-of-green (SOG) and screen-of-green (SCROG) training and trellising methods. In soil types, fertilizers, and feeds similar advances were made. The plant itself was constantly refined, bred, and cloned into many new

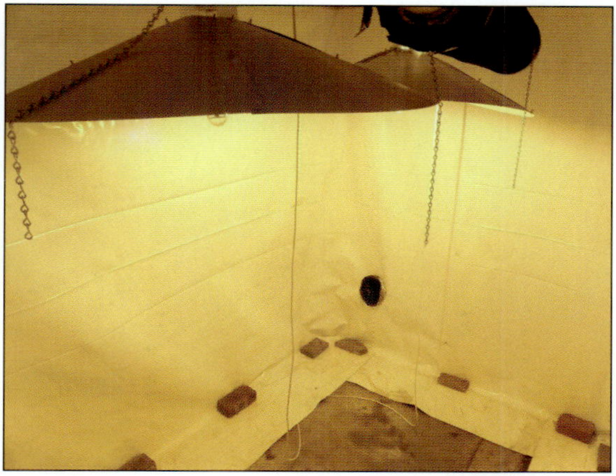

Left and this page: From seed to seedlings in a grow room under suitable lights

The growing cycle continues.

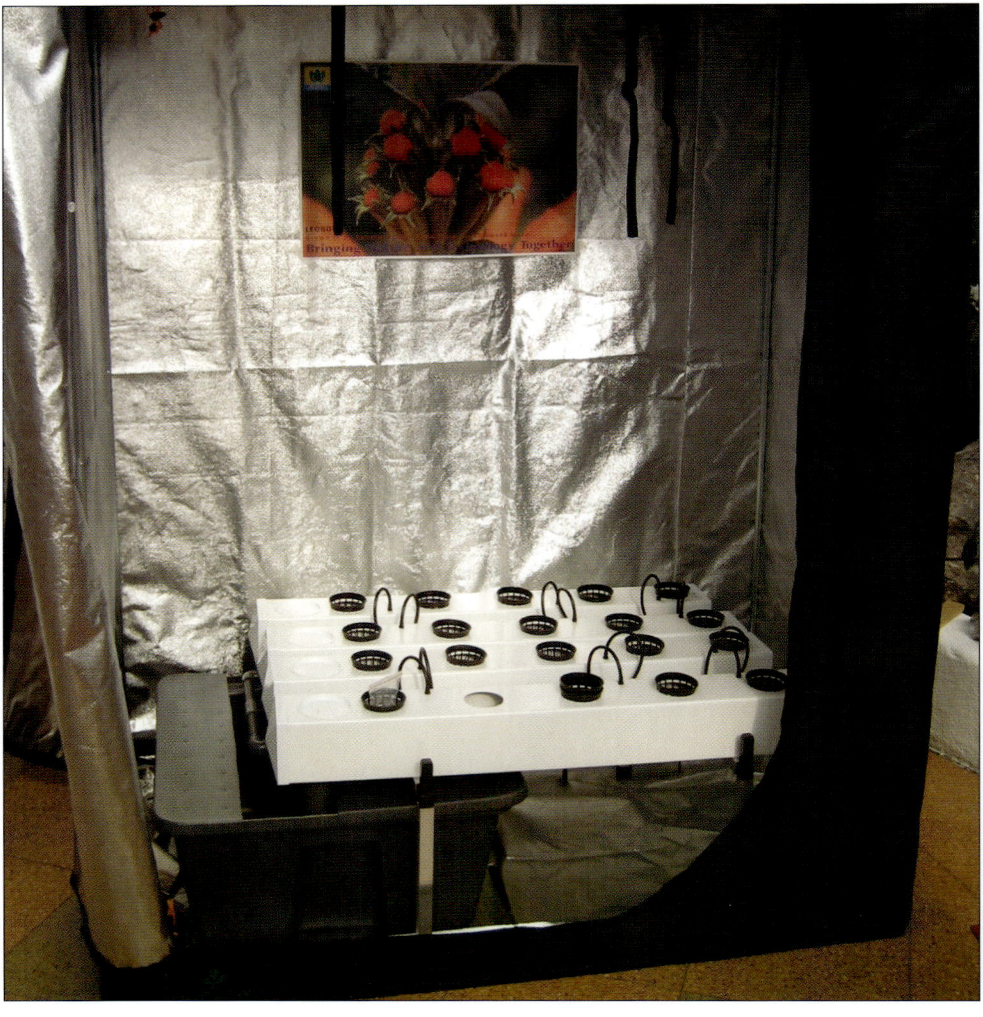
Bringing
Together

strains for size, taste, yield, potency, and appearance, primarily of the dried buds of the female flowers. This is accomplished by buying specifically guaranteed female seeds or plants, or by monitoring the plants as they grow and removing the males (which produce little THC) before they pollinate the females. As the female plants flower, their buds produce trichomes–little hairs that are encrusted with crystals in which resides the psychoactive THC. What is then harvested is called *sinsemilla* (without seed), also known as "skunk"–originally the name of a specific strain that has now become the generic term for this especially strong and smelly kind of weed. Its distinctive and easily detectable smell has led to the use of further "tech" solutions: ventilation extractors fitted with carbon filters and ozone machines that negate its odor. Air-ported police thermal imagers that can identify large heat sources are countered with more energy-efficient lighting systems and masking strategies.

For the last 40 years, California and Holland have led the development of homegrown systems and seed selection. The Internet, knowing no national boundaries, has further enabled the exchange of information and resources to advance the growth of both recreational and medicinal use. A truly vast array of growing and lighting equipment is now available, along with seed selection catalogs listing an endless variety of hybrids. A library of Internet and published manuals guide the grower from start to finish and the paraphernalia of vaporizers, bongs, pipes, rolling papers, and other equipment has become an industry in itself.

Above: ACE-mix skunk bud with reddish trichome
Left: Grow tent with hydroponic system installed

Left and right: The heavy buds of "Chronic" skunk ready for imminent harvesting. Sodium lights account for the yellow cast on this and similar images.

Above: Pink hydroponic pellets for female cultivators?
Right: "White Rhino" skunk seed pack
Far right: Amsterdam cannabis seed sales stall

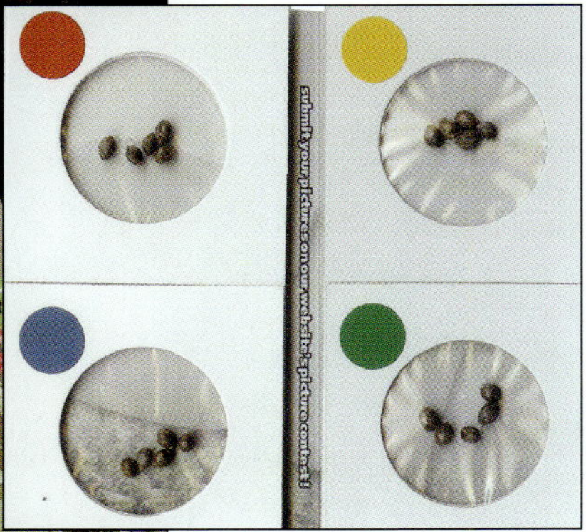

Top and above: A range of seeds on sale in Amsterdam

This page: Top left: "White Rhino"
Bottom left: "Big Buddha Cheese"
Above: "White Russian"

Opposite page: Left: A bulging "Chronic" bud
Right: Home grow room with plants, lights, and carbon air filter

Left: "Blue Cheese" room under lights

Right: "Northern Lights"

Right and far right: Two views of "Northern Lights" growing in Spain

Right and far right: Two more views of "Chronic" superbud

Above: "Blue Hammer"
Right: "Superbud"

Overleaf: Purple bud in the rain

MEDICAL ADVANCES

A substantial body of respectable, acknowledged evidence has accumulated for the beneficial effects of cannabis in treating a wide range of conditions including AIDS, glaucoma, Parkinson's, cancer, motor neurone disease, multiple scelerosis, and depression as well as serving as a general analgesic and anti-spasmodic. This research has spurred the development of synthetic forms of cannabinoids sold as prescription drugs, such as Dronabinol and Nabilone. The medical authorities attempt to remove or control the psychoactive element of the drug in order to destroy its recreational capability.

This medical use has been primarily driven by insistent private use, implying proof of its efficacy. Normal law-abiding citizens with medical problems that have not been solved by the pharmaceutical industry have sought relief through cannabis despite the continuing legal prohibitions; some in the medical establishment have acknowledged these benefits in research and release of prescriptive remedies.

State governments have also begun to take notice and the process of decriminalization has begun. Although at the Federal level medical cannabis use remains a criminal offence, there are important legal differences between the federal and state governments, with further guidelines being enacted in order to relinquish the initiative to the state level.

U.S. Attorney General Eric Holder has said: "It will not be a priority to use federal resources to prosecute patients with serious illnesses or their caregivers who are complying with state laws on medical marijuana, but we will not tolerate drug traffickers who hide behind claims of compliance with state law to mask activities that are clearly illegal."

A contradiction exists in his press-prepared

Official jar for prescription marijuana (left) and dispensary (above)

words. To date, 16 states have decriminalized the medical use of marijuana: Arizona, Alaska, California, Colorado, Delaware, Hawaii, Maine, Michigan, Montana, Nevada, New Jersey, New Mexico, Oregon, Rhode Island, Vermont, and Washington. Dispensaries in these states are licensed to sell marijuana to legally entitled people for medical use. Marijuana is also in many other products, such as cakes, chocolate bars, nachos, and other baked foods, as an alternative mode of ingestion for those who do not smoke. Public opinion increasingly favors legalization, and state authorities are becoming more aware of its earning potential. Marijuana is one of the top-selling

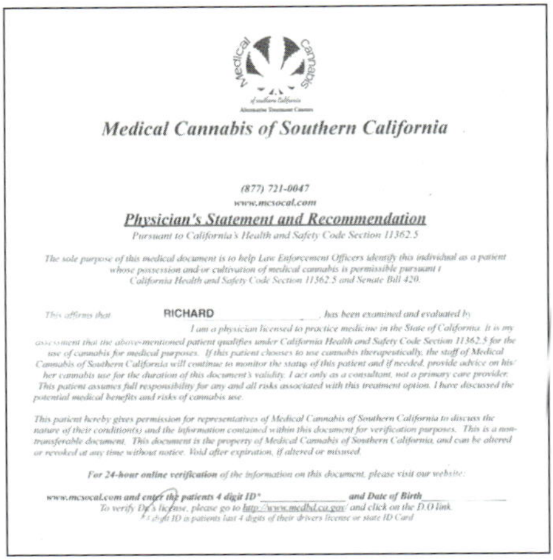

Medical Cannabis of Southern California

(877) 721-0047
www.mcsocal.com

Physician's Statement and Recommendation

Pursuant to California's Health and Safety Code Section 11362.5

The sole purpose of this medical document is to help Law Enforcement Officers identify this individual as a patient whose possession and/or cultivation of medical cannabis is permissible pursuant t California Health and Safety Code Section 11362.5 and Senate Bill 420.

This affirms that RICHARD __________, *has been examined and evaluated by*

I am a physician licensed to practice medicine in the State of California. It is my assessment that the above-mentioned patient qualifies under California Health and Safety Code Section 11362.5 for the use of cannabis for medical purposes. If this patient chooses to use cannabis therapeutically, the staff of Medical Cannabis of Southern California will continue to monitor the status of this patient and if needed, provide advice on his/her cannabis use for the duration of this document's validity. I act only as a consultant, not a primary care provider. This patient assumes full responsibility for any and all risks associated with this treatment option. I have discussed the potential medical benefits and risks of cannabis use.

This patient hereby gives permission for representatives of Medical Cannabis of Southern California to discuss the nature of their condition(s) and the information contained within this document for verification purposes. This is a non-transferable document. This document is the property of Medical Cannabis of Southern California, and can be altered or revoked at any time without notice. Void after expiration, if altered or misused.

For 24-hour online verification *of the information on this document, please visit our website:*

www.mcsocal.com and enter the patients 4 digit ID*________ and Date of Birth________
To verify Dr.'s license, please go to http://www.medbd.ca.gov/ and click on the D.O link
**4 digit ID is patients last 4 digits of their drivers license or state ID Card*

products in the U.S., with estimated annual sales of over $10 billion. The tax revenues to be earned if the drug were fully legalized and regulated are quite substantial, as is the money currently spent on failed enforcement.

Some, such as the Establishment *bete noire* J. Orlin Grabbe in *The Laissez Faire City Times*, have argued: **"The function of the Drug War is to create the Drug Crisis."** The Drug Crisis involves billions of dollars in hidden cash flow. Addicted to this flow of money are law enforcement agencies, drug producers and distributors, covert agencies who use it as a source of black funding, and politicians and bankers who are hired to protect the drug revenues. Addiction to drug revenues requires that the drug war be fought so as to be lost. Failure thus becomes the criterion of success.

Left: Medical dispensary sign
Left and below: Physician's letter of recommendation and Californian cannabis-user's Medical Cannabis Identification Card

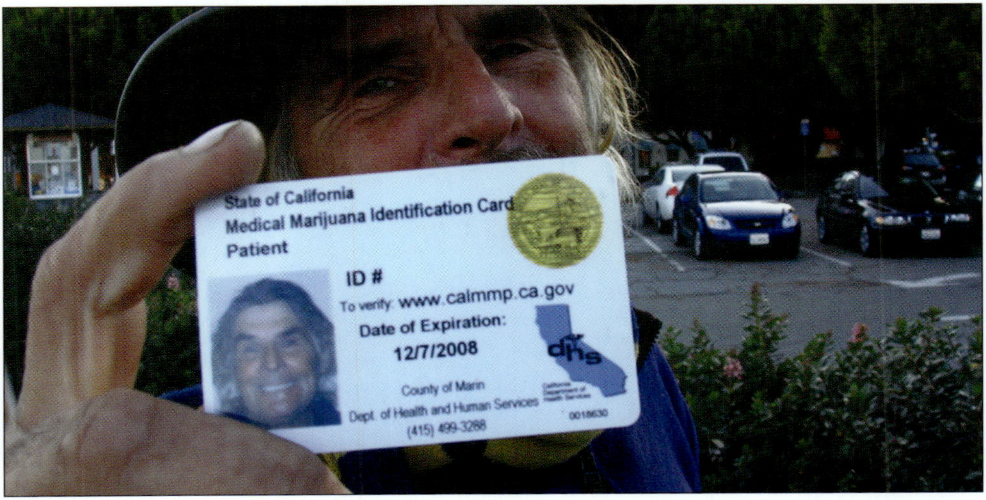

Left: American medical hashish: A-grade OG Kush trichome

Above: American medical hashish: Grape Kush bubble hash, purified and cured trichomes from Grape Kush C. indica

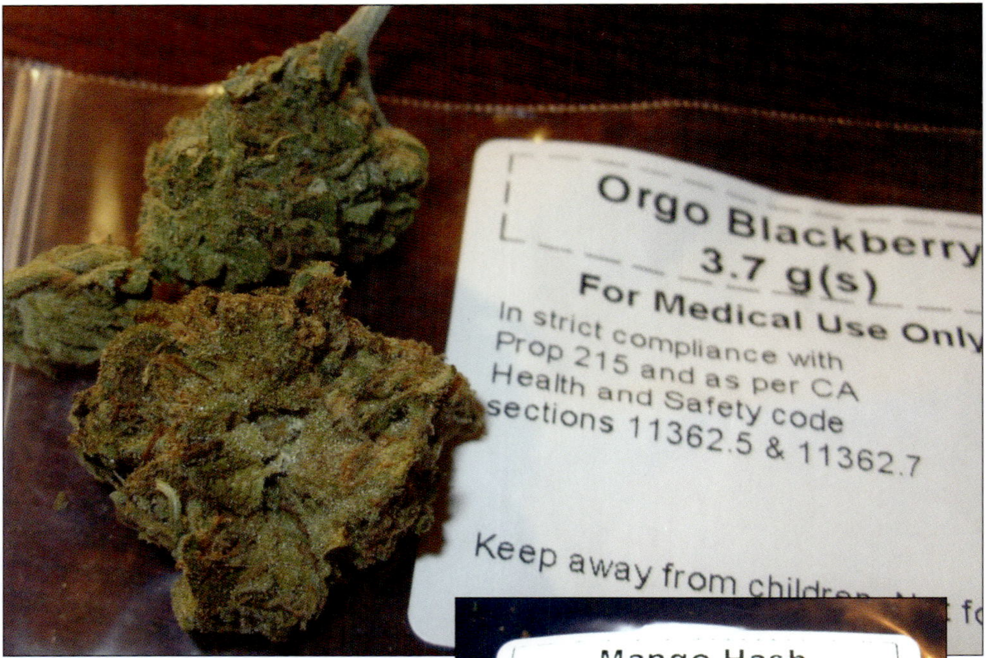

Above: Organic C. indica *purchased from a Medical Cannabis dispensary per California Proposition 215/Amendment #420/ Health and Safety code 11362.5 and 11362.7*

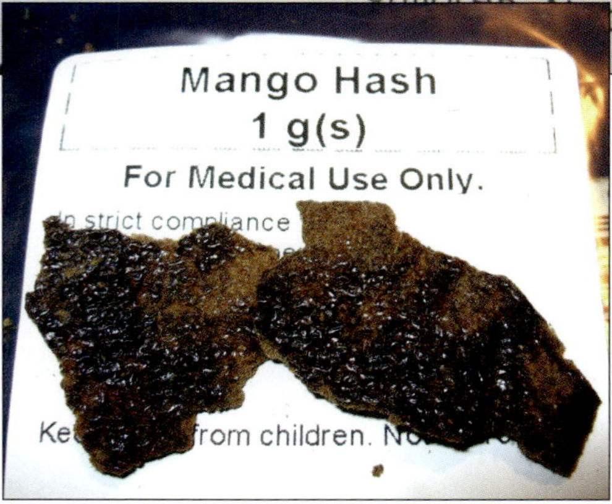

Right: One gram of isolated and concentrated 76-micron cannabis trichomes obtained from "Master Mango" by the Bubble Bag method. Note the dark, glossy layer where the purest resin has settled down onto the curing paper

“Ganesh” hashish wafer— a commercially prepared C. indica *pressed bubble hash*

LEGALITY IN THE U.S.

In the U.S., the current position of the Drug Enforcement Agency is that "marijuana is properly categorized under Schedule 1 of the Controlled Substances Act (CSA), 21 U.S.C. § 801, et seq." The clear weight of the currently available evidence supports this classification, including evidence that smoked marijuana has a high potential for abuse, has no accepted medicinal value in treatment in the United States, and evidence that there is a general lack of accepted safety for its use even under medical supervision."

At the federal level, possession of marijuana is punishable by up to one year in jail and a minimum fine of $1,000 for a first conviction. For a second conviction, the penalties increase to a 15-day mandatory minimum sentence with a maximum of two years in prison and a fine of up to $2,500. Subsequent convictions carry a 90-day mandatory minimum sentence and a maximum of up to three years in prison and a fine of up to $5,000.

Distribution of a small amount of marijuana, for no remuneration, is treated as possession. Manufacture or distribution of less than 50 plants or 50 kilograms (kg.) of marijuana is punishable by up to five years in prison and a fine of up to $250,000. For 50–99 plants or 50–99 kg. the penalty increases to not more than 20 years in prison and a fine of up to $1 million if an individual, $5 million if other than an individual, for the first offense. Manufacture or distribution of 100–999 plants or 100–999 kg. carries a penalty of 5–40 years in prison and a fine of $2–$5 million. For 1,000 plants or 1,000 kg. or more, the penalty increases to from 10 years to life in prison and a fine of $4–$10 million.

Distribution of greater than five grams of marijuana to a minor under the age of 21 doubles the possible penalties. Distribution within 1,000 feet (ft.) of a school, playground, public housing, or within 100 ft. of a youth center, public pool or video arcade also doubles the possible penalties.

The sale of paraphernalia is punishable by up to three years in prison.

The death sentence can be given to a defendant who is found guilty of manufacturing, importing, or distributing a controlled substance if the act was committed as part of a continuing criminal enterprise, but only if the defendant is the principal administrator, organizer, or leader of the enterprise and the quantity of the controlled substance is 60,000 kg. or more of a mixture or substance containing a detectable amount of marijuana, or 60,000 or more marijuana plants, or if the enterprise received more than $20 million in gross receipts during any 12-month period of its existence.

Right: C. sativa *hemp plants*

Left and right: Outdoor homegrown plants

State	Possession	Sale	Cultivation	Medical Use
Alabama	Misdemeanor	Felony	Felony	Illegal
Alaska	Decriminalized	Felony	Felony	Decriminalized
Arizona	Felony	Felony	Felony	Decriminalized
Arkansas	Misdemeanor	Felony	Felony	Illegal
California	Decriminalized	Felony	Felony	Decriminalized
Colorado	Decriminalized	Decriminalized	Felony	Decriminalized
Connecticut	Decriminalized	Felony	Felony	Decriminalized
Delaware	Misdemeanor	Felony	Felony	Decriminalized
Florida	Misdemeanor	Felony	Felony	Illegal
Georgia	Misdemeanor	Felony	Felony	Illegal
Hawaii	Misdemeanor	Felony	Felony	Decriminalized
Idaho	Misdemeanor	Felony	Felony	Illegal
Illinois	Misdemeanor	Felony	Felony	Illegal
Indiana	Misdemeanor	Felony	Felony	Illegal
Iowa	Misdemeanor	Felony	Felony	Illegal
Kansas	Misdemeanor	Felony	Felony	Illegal
Kentucky	Misdemeanor	Felony	Felony	Illegal
Louisiana	Misdemeanor	Felony	Felony	Illegal
Maine	Decriminalized	Felony	Felony	Decriminalized
Maryland	Misdemeanor	Felony	Felony	Illegal
Massachusetts	Decriminalized	Felony	Felony	Illegal
Michigan	Misdemeanor	Felony	Felony	Decriminalized
Minnesota	Decriminalized	Felony	Felony	Illegal
Mississippi	Decriminalized	Felony	Felony	Decriminalized
Missouri	Misdemeanor	Felony	Felony	Illegal
Montana	Misdemeanor	Felony	Felony	Decriminalized
Nebraska	Decriminalized	Felony	Felony	Decriminalized
Nevada	Decriminalized	Felony	Felony	Decriminalized
New Hampshire	Misdemeanor	Felony	Felony	Decriminalized

State	Possession	Sale	Cultivation	Medical Use
New Jersey	Misdemeanor	Felony	Felony	Decriminalized
New Mexico	Misdemeanor	Felony	Felony	Decriminalized
New York	Decriminalized	Felony	Felony	Decriminalized
North Carolina	Decriminalized	Felony	Felony	Illegal
North Dakota	Misdemeanor	Felony	Felony	Illegal
Ohio	Decriminalized	Felony	Felony	Decriminalized
Oklahoma	Misdemeanor	Felony	Felony	Illegal
Oregon	Decriminalized	Felony	Felony	Decriminalized
Pennsylvania	Misdemeanor	Felony	Felony	Illegal
Rhode Island	Misdemeanor	Felony	Felony	Decriminalized
South Carolina	Misdemeanor	Felony	Felony	Illegal
South Dakota	Misdemeanor	Felony	Felony	Illegal
Tennessee	Misdemeanor	Felony	Felony	Illegal
Texas	Misdemeanor	Felony	Felony	Illegal
Utah	Misdemeanor	Felony	Felony	Illegal
Vermont	Misdemeanor	Felony	Felony	Decriminalized
Virginia	Misdemeanor	Felony	Felony	Illegal
Washington	Decriminalized	Decriminalized	Felony	Decriminalized
West Virginia	Misdemeanor	Felony	Felony	Illegal
Wisconsin	Misdemeanor	Felony	Felony	Illegal
Wyoming	Misdemeanor	Felony	Felony	Illegal
Washington DC	Misdemeanor	Felony	Felony	Decriminalized

Left: Emile Bernard's La fumeuse de Haschisch *was painted in 1900 during the artist's 10-year stay in Cairo. It shows a woman with a nose ring and a narghile. Today it hangs in the Musée d'Orsay.*

Below: Smoking dope

Right: Sharing a joint, 1969. Today's counter culture owes much to 1960s California.

Left: Vietnamese traditional pipes with rustic tobacco

Right: Dutch coffee shop and bong

Far right: Volcano vaporizer filling up with "Kush" vapor

Volcano

Above and right: Two of a sequence of photographs taken by photographer Marc St. Gil in cedar woods near San Antonio in the 1970s

Above: Amsterdam was the mecca for European cannabis users. These three hash cakes were made with hashish from different Amsterdam coffee shops.

Left: Amsterdam store window displaying various medical cannabis, hemp food, and other types of Cannabis products

Right: The Hash, Marijuana, and Hemp Museum opened in Amsterdam in 1985

SENSI SEEDS
WWW.SENSISEEDS.COM
Hash Marihuana
& Hemp Museum
WWW.HASHMUSEUM.COM

blowverbod
wegens overlast in de buurt
boete €50,- art. 2.8 lid 4 APV
% ALC
alcoholverbod
art. 2.8 lid 2 APV
uitgezonderd terras

PIC
PIC

UP THE PIECES
UP The PIECES
ALWAYS OPEN COFFEESHOP
FFEESHOP
OPEN
RCONDITIONED
OPEN
UP the PIECES
7

Previous spread, above, and right: In 1990, 1,450 coffee shops were located in The Netherlands–400 in Amsterdam, the others spread over the country, mainly in the bigger cities and the border areas with Belgium and Germany. A ban on coffee-shop sales of cannabis to foreigners is now being implemented.

OPEN
COFFEESHOP

LEGALITY AROUND THE WORLD

The Netherlands is the unofficial global headquarters of undeterred, out-in-the-open cannabis use, although it is still technically illegal, and coffee-shop sales to tourists have been banned.

Laws vary among nations to a great degree. It is illegal to use, possess, grow or sell cannabis in Australia, but penalties differ for each state or territory. Penalties for possession or use of marijuana in Japan are severe, and convicted offenders can expect long jail sentences and fines. Since 2009, Mexico has decriminalized personal-use possession for up to five grams of marijuana (the equivalent of four marijuana cigarettes); the change only recognized the country's longstanding practice of not prosecuting people caught with small amounts of drugs. The use, possession, cultivation, and supply of cannabis remains illegal (but often unenforced) in the Republic of South Africa, the first country to enact anti-cannabis legislation at the turn of the 20th century in 1908. In Switzerland, as of Jaunary 1, 2012, home-grown cannabis is legal for up to four plants per grower. The sale of cannabis, however, remains illegal.

The medicinal use of cannabis is legal in Canada, Israel, and the Czech Republic in addition to 16 U.S. states.

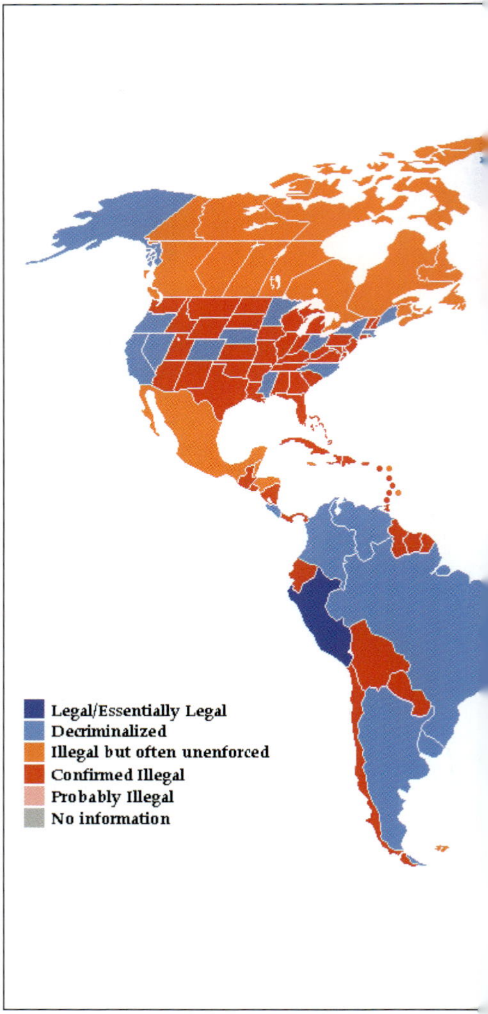

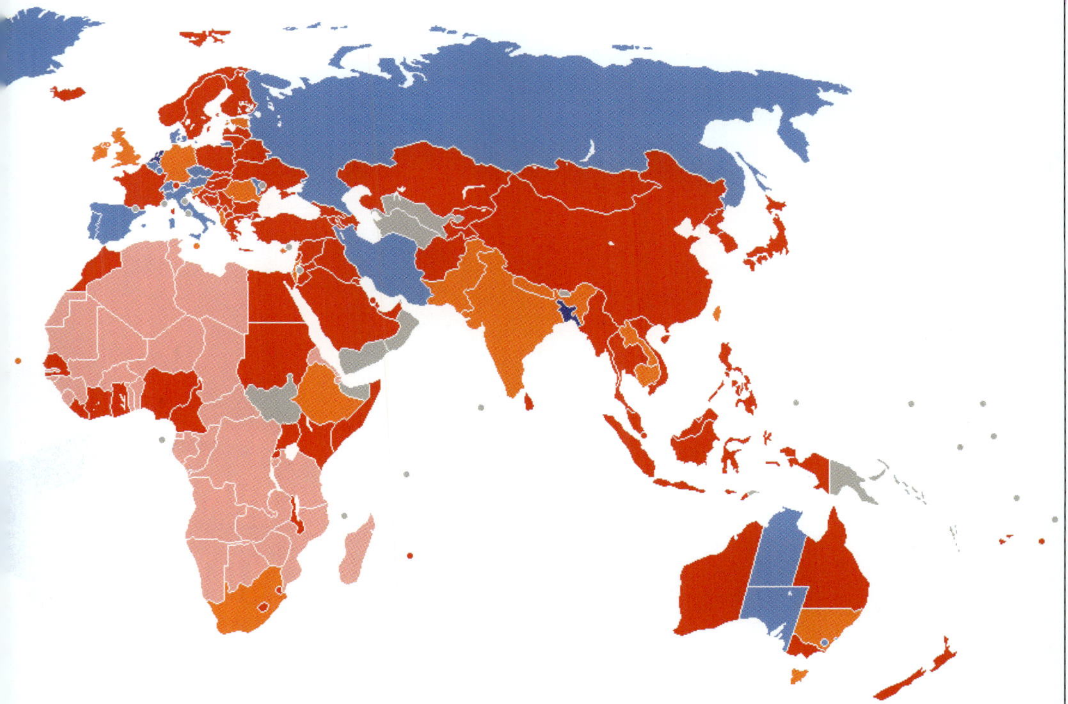

USEFUL WEBSITES/ ADDRESSES

ASA (http://safeaccessnow.org/), or Americans for Safe Access, is the largest national member-based organization of patients, medical professionals, scientists, and concerned citizens promoting safe and legal access to cannabis for therapeutic use and research. ASA is dedicated to meeting the immediate needs of medical cannabis patients and their providers while creating a movement to promote safe and legal access. ASA works to overcome political and legal barriers by creating policies that improve access to medical cannabis for patients and researchers by engaging in a multi-faceted strategy that incorporates public education, impact litigation, grassroots development and advocacy, media campaigns, and direct support services.

CANNABIS FREEDOM ACTIVIST NETWORK (http://www.freedomactivist.net/cannabis.html) provides information, links and other resources to facilitate communication, interaction, cooperation, and synergy among those fighting for cannabis legalization, personal freedom, and the right to individual choice.

HIGH TIMES (http://hightimes.com/) is a well-known New York-based monthly magazine and website with a huge archive devoted to the legalization and enjoyment of cannabis.

NORML (http://norml.org), or National Organization for the Reform of Marijuana Laws, was founded in the U.S. in the 1970s. NORML campaigns for the removal of all criminal penalties for the private possession and responsible use of marijuana by adults, including the cultivation for personal use, and the casual nonprofit transfers of small amounts, as well as the development of a legally controlled market for cannabis. NORML and the NORML Foundation support all people fighting prosecution under marijuana laws and those working to legalize marijuana.

INDEX